The Cornfield Maze Mystery

A Story About Food, Farming, and Discovery

By David Upthegrove

For All of the Educators

Copyright © 2025 David Upthegrove

ISBN: 979-8-9934904-3-4

First Edition

Published 2025

The Cornfield Maze Mystery

Introduction

This book was written to help young readers learn about the food we eat, especially corn and to spark curiosity about where food comes from and how it affects our bodies. Corn is in a lot of food products many of us eat every day, so it makes a great place to start asking questions: How does corn grow? What happens on farms? What do farmers use to protect crops and why might that matter to us?

We'll explore a couple of examples; a weed killer called glyphosate, and genetically modified corn. This information is not to frighten you, but to show how chemicals used in farming can sometimes end up in soil, water, and food. Scientists are still learning exactly what long-term, small amounts of these chemicals do to our bodies' and whether bio engineered foods impact health. That's why asking questions, reading carefully, and learning how food is grown are important skills. Knowledge is power: the more we understand the better choices we can make for our health.

The language is simple; the ideas are explained in short chapters. My hope is to inspire curiosity, not fear, to encourage you to explore, experiment, and discover how a clean food supply helps bodies grow strong and minds stay sharp.

This story begins with three curious kids, a cornfield maze, and questions?

The goal of this book is not to frighten readers but to encourage curiosity and thoughtful questions. Along the way they learn how food moves from farm to table and how communities can work together to make change.

If this book raises questions you'd like to explore with children, there is an educator/parent guide with activities, sources, and classroom experiments available.

Table of Contents

A comprehensive Educator and Parent Guide with detailed lesson plans, experiments, activities, assessments, and teaching resources is available as a separate companion book.

Chapter 1 - Harvest BBQ at Uncle Rays

They were at Uncle Ray's for the last-blast-of-summer family barbecue. On this farm, the corn grew in tidy rows that seemed to stretch forever—perfect for a maze. From the hilltop, Uncle Ray's farm looked larger than on the map: green rows of corn marching like a small, patient army. The red barn that leaned just enough to seem friendly and the maze made of paths weaving through tall cornstalks forming a living puzzle.

Mary, Ana and Ben were excited to be at the farm. The family BBQ was a long standing tradition held every year before the corn harvest. The long gravel driveway smelled like sun-warmed dust and freshly cut grass. There was light smoke and the smell of BBQ drifting from behind the house.

Mary, Ana, and Ben tumbled from the pickup and ran across the yard. They were the Corn Crew: Mary tightened the laces of her sneakers and grinned. At eleven, she liked facts the way some kids liked video games—she collected them and thought about how they fit together. Beside her, Ana, also eleven, swung a ribbon in her hair and laughed at a bee that would not be shooed.

Ana's family had moved into town two years earlier and she loved stories about where food came from. Ben, a ten-year-old with a quick smile and quicker questions, carried a small notebook. He liked to write down things he wanted to remember.

Uncle Joe greeted the children and handed each of them a paper map with bold letters: FIND THE FLAG. He warned them to mind the rows. It sounded like a game, and at first it was. "It's like a labyrinth," Ben said, ducking between stalks that were taller than any of them.

"Let's try to get to the pond," Mary suggested. "First one there names the next game."

They ran, ducking and spinning between green walls. The corn leaves rustled at their shoulders like the pages of a book. The sun slid toward the edge of the sky, and the air smelled faintly of smoke from the barbeque.

They raced between the towering stalks, but then Mary noticed something strange: a patch where the leaves looked dull and powdery, where kernels were smaller and the corn silk had a crust of sticky droplets.

A crooked wooden sign read: SPRAYED—DO NOT ENTER.
The maze shifted from a game to a question. The Corn Crew decided to be scientists. After they'd played until their legs ached, Uncle Ray found them with a sun-faded cap in his hand and a smile on his face.

"You explorers ready for a farm tour?" he asked. Uncle Ray was gentle and patient. He owned the fields that rolled behind the house and had farmed the land for longer than any of them had been alive.

"Yes!" they answered together. They would ask Uncle Ray what he had sprayed and why.

Chapter 2 - Machines and Seeds

Uncle Ray led them across a track of gravel to where the big machines were stored between jobs. He spoke with pride and simple honesty.

"This is a tractor," he said, patting the big tire on a blue machine. "This one pulls things—planters at planting, grain carts at harvest, trailers to move soil or seed."

Mary looked closely at a long, narrow machine with rows of metal discs. "What's that?"

"That's a row-crop planter," Uncle Ray said. "It drops seeds precisely into the soil when spring comes. Farmers want the plants evenly spaced so each one gets enough sun and nutrients."

They passed a broad machine with long arms and nozzles.

"That's a boom sprayer," Uncle Ray explained. "It sprays herbicide,(pronounced as "HER-buh-syd.") or fertilizer across the field. Sometimes people use airplanes—crop dusters—to spray, but for big corn fields we usually use these or big self-propelled sprayers."

Ben's eyebrows shot up. "So the sprayer uses...poison?"

Uncle Ray shook his head. "Herbicides are chemicals that target plants we don't want—like weeds. They're made to be effective against certain plant processes. We use them carefully and follow rules about how much, when, and how to apply."

Near the storage sheds they saw big tanks and labeled drums. A sign read: Authorized Personnel Only—Chemical Storage.

"We keep chemicals locked up for safety," Uncle Ray said. "And we wear protective equipment when handling them."

Ana peered at a narrow bag labeled with small print. "What about the seeds? You said they were special earlier when Mary asked."

Uncle Ray smiled. "Good question. Seed companies develop varieties that are bred or genetically modified to tolerate certain herbicides or to resist insects. That means we can control weeds without hurting the crop, and insects won't eat the corn the way they used to."

Mary tilted her head. "So the corn itself is changed?"

"Yes," Uncle Ray said. "Some seeds are hybrid—bred the old-fashioned way. Others are genetically modified. That means scientists have inserted specific genes so the plant can make a protein that protects it from a pest—like Bt corn does—or so the plant won't be harmed by a certain herbicide."

He explained that the right choice meant reading labels, choosing timings, and understanding trade-offs. Sometimes spraying too close to the plant or at the wrong time could stress the corn. Sometimes banks and buyers asked for certain yields and varieties that pushed farmers toward quick fixes.

The kids learned that farming mixed science with money, patience, and worry. Uncle Ray agreed to help them test the soil so they could learn more.

Ben scribbled in his notebook, more questions already forming. He noticed, in the distance, a field being gently misted by a sprayer. The three friends watched with a mix of awe and worry.

Chapter 3 - Questions at the Edge of the Field

As the tour wound down, Uncle Ray led them past silos and grain bins—big metal cylinders that would hold the yellow kernels once the harvest came in.

"After the combine runs through, the corn goes to the grain cart, then into trucks," Uncle Ray said, pointing. "Then to dryers and finally to the elevator or processing plant. From there it could become food ingredients, animal feed, or even ethanol."

Ana stopped. "If you spray herbicides, does any of that end up in the corn?"

Uncle Ray's face grew serious. "Most of these herbicides target pathways in plants. The corn varieties that are tolerant aren't killed by them. Processing and cleaning reduce residues, and there are legal limits called maximum residue limits that regulators set. But people are right to ask about what's left and whether it matters for people and the environment."

Ben added, "My cousin said some types of corn have a protein that kills bugs."

"That's Bt," Uncle Ray said. "It comes from a natural bacterium, Bacillus thuringiensis. (pronounced as "buh-SIL-us thur-in-JEN-sis.") Scientists can put that gene into the corn so it makes a protein harmful to certain insects but safe for people. It helps farmers reduce insecticide sprays."

The kids thanked Uncle Ray and headed back to the house, their minds buzzing with new questions. The following Monday, school would start.

Chapter 4 - Back-to-School and the Science Fair

On Monday the three of them met at the brick steps of Lincoln Elementary, cheeks flushed from their parents' goodbyes.

Inside class, Ms. Patel—bright and clear-voiced—announced, "This year's science fair will include group projects. Teams of three will research and make displays. You'll have four weeks. Use your homework time to plan."

Mary exchanged a look with Ana and Ben. Perfect timing, she thought. Their end-of-summer curiosity had just found a deadline.

"Team?" Mary mouthed.

Ana and Ben nodded.

They signed up as a team of three before the bell rang. Ms. Patel smiled when they told her they wanted to study corn production from seed to table. "That sounds fantastic," she said. "Make sure to include background, experiments or fieldwork, and a clear conclusion. And remember to be fair and balanced in how you present evidence."

The project was officially on. They needed a plan. Their first homework assignment was to brainstorm questions and decide how to investigate.

They made a list:

- How is corn planted, grown, and harvested?

- What machines are used and when?

- What is genetically modified corn? Why is it used?

- How do herbicides and insecticides work?

- Does any chemical end up in food, and could that affect people?

- How many everyday foods contain corn or corn-derived ingredients?

They split tasks: Mary would talk to Uncle Ray again about another visit to the farm to take photos of machines and gather more information; Ana would talk with her family and keep a daily note about her mother's stomach trouble—just in case; Ben would talk to his mom and ask if they could speak with the family doctor, Dr. Lopez, and help arrange a field trip to a processing plant.

Their first stop was obvious—Uncle Ray's harvest.

Chapter 5 - Harvest Day at the Field

Two weeks later they arrived at the farm before dawn. The air hummed with activity. A combine harvester—big and boxy with a rotating header—moved like a steady animal through the rows. It snapped the ears of corn from the stalks, separated kernels from husks, and sent the grain into a waiting grain cart.

"Watch how the combine feeds into the grain cart," Uncle Ray said. "When the cart's full, it unloads into trucks which go to the grain dryer or straight to the elevator."

Mary took careful photos. Ana recorded video. Ben counted trucks like a game.

They watched as a tractor pulled a chisel plow through a different field, turning the soil in preparation for next spring. A seed tender—Uncle Ray explained—carried bags of seed near the planter to refill it when the planter ran out.

"Everything's timed," Uncle Ray said. "Planting in spring, spraying when weeds are young, scouting for insects through the summer, harvest in late summer or fall. We try to use tools that help us be efficient and reduce waste."

As the sun rose higher Uncle Ray drove them to the edge of the field where workers were preparing a boom sprayer to move to another section. Ben noticed the label on a tank, glyphosate, (pronounced "gly-fuh-sate").He wrote the name in his notebook with bold, neat letters.

The Corn Crew learned to look with new eyes about where our food comes from and how it gets to grocery stores. Starting from the ground, soil was not just dirt. It was a neighborhood where worms dug tunnels, microbes munched bits of old leaves, and roots whispered. They learned that soil held water like a sponge and nutrients like a tiny grocery store for plants.

For the first experiment, they used a trowel and an old yellow bucket. They took samples from many places, separated them, and labeled the bags. In a small notebook they recorded color, smell, and texture. They compared darker, crumbly soil to pale, dusty patches and wondered what made the difference.

They learned healthy soil had life and structure. When soil lost its organic matter and worms disappeared, it could not hold as much water and food for plants. Planting cover crops, using compost, and crop rotation could help maintain soil health. The Corn Crew began to see soil as an important part of farming.

Ana excitedly mentioned the school field trip to the processing mill the following week.

Chapter 6 - Processing Plant Visit

Welcome to
Heartland Processing

Ben's mom and Ms. Patel had arranged a school field trip to the nearby Heartland Processing Plant. The three friends bounced excitedly in their seats on the bus ride over, notebooks ready.

The building was huge—taller than their school gymnasium, with loading docks and grain silos connected by metal tubes and conveyor systems. A friendly guide named Ms. Rodriguez greeted them at the entrance.

"Welcome to Heartland Processing!" she said with a smile. "Today you'll see how corn kernels become many different products. Safety first—everyone stay behind the yellow lines."

The lobby had large posters showing the journey of corn:

- Whole Kernel → Cleaning → Steeping → Milling → Separation → Products

- Products shown: Corn starch, corn syrup, corn oil, corn meal, corn gluten

"We process about 50,000 bushels of corn every day," Ms. Rodriguez explained as they walked down a hallway with windows looking into the plant floor. "That's enough to fill about 70 large grain trucks."

The first stop was the receiving area. Through the glass, students could see trucks backing up to unload golden kernels that cascaded like waterfalls onto conveyor systems.

"Before we accept corn, we test it," Ms. Rodriguez said, holding up a small sample container. "We check moisture content, test weight, and look for foreign materials or damage. High-quality corn makes high-quality products."

Mary raised her hand. "Do you test for herbicide residues?"

Ms. Rodriguez nodded. "Good question! Yes, we do periodic testing. Regulatory agencies like the FDA set maximum residue limits, or MRLs, which are safety thresholds. Most corn we receive is well below those limits. It is an important part of quality control."

The class moved to the cleaning station. Giant rotating screens and air blowers removed dust, broken kernels, and bits of cobs. The clean kernels moved on to the next stage.

Wet Milling vs. Dry Milling

"There are two main ways to process corn," Ms. Rodriguez explained. "Wet milling and dry milling. We do wet milling here."

She led them to a row of large steel tanks filled with water. "In wet milling, we soak the kernels in warm water with a little sulfur dioxide for 24 to 48 hours. This is called steeping. It softens the kernel and helps us separate the different parts."

Ben wrote furiously in his notebook. Ana took photos of the tanks with their bubbling contents.

"After steeping, the kernels are ground up and separated into four main components," Ms. Rodriguez continued, pointing to a diagram:

1. Germ (the part that would sprout) → becomes corn oil

2. Fiber (outer hull) → becomes animal feed or fiber products

3. Gluten (protein) → becomes animal feed or additives

4. Starch (the biggest part) → becomes corn starch, corn syrup, and other sweeteners

Through another set of windows, they watched centrifuges spinning at high speeds, separating the germ from the rest. The germ floated because it contained oil—lighter than water.

"The germ goes through an extraction process to get the oil out," Ms. Rodriguez said. "Corn oil is used for cooking, in salad dressings, and even in some non-food products like soaps."

11

The next station showed the fiber being dried and pressed into animal feed pellets. The gluten—bright yellow and protein-rich—was also being processed for animal feed.

But the starch was the star of the show. Ms. Rodriguez led them to a control room overlooking massive stainless steel tanks.

"Corn starch can be sold for cooking and industrial uses," she explained. "Or we can process it further using enzymes to create sweeteners and other products."

Mary raised her hand. "Where does all this corn come from? Is it genetically modified?"

Ms. Rodriguez nodded. "Good question! Most of the corn we process is genetically modified. About 90% of corn grown in the United States is GMO—that means scientists have changed its DNA to give it specific traits."

"What kind of traits?" Ben asked, writing quickly in his notebook.

"Two main types," Ms. Rodriguez explained. "Some GMO corn produces a protein called Bt toxin that kills certain insect pests. This reduces the need for insecticide spraying. Other GMO corn is herbicide-tolerant—usually engineered to survive glyphosate, a common weed killer. Farmers can spray the whole field, and the weeds die but the corn doesn't."

Ana looked concerned. "Is that safe? The herbicides, I mean?"

"That's an important question scientists and regulators continue to study," Ms. Rodriguez said. "Government agencies set safety limits for herbicide residues in food. But there are debates about long-term effects, especially on beneficial insects, soil microbes, and water quality."

She led them to the quality control laboratory. Scientists in white coats worked at benches with microscopes, test tubes, and computer screens.

"We test every batch," one scientist explained. "We check for quality, safety, and chemical residues."

"Do you test for herbicide residues?" Mary asked.

The scientist nodded. "Yes, periodically. The washing and steeping process removes most water-soluble chemicals like glyphosate. But we do testing to ensure we meet federal safety standards—called Maximum Residue Limits, or MRLs."

Ben leaned forward. "What if the residues are higher than allowed?"

"Then that batch can't be used for food products," the scientist replied. "Food safety is our top priority. But it's also why understanding where our corn comes from matters. Different farming practices lead to different residue levels."

Mary looked at the processing equipment with new eyes. "So when we buy corn products at the store, we might not know if they came from GMO corn or if there are herbicide residues?"

"In the United States, most corn-based products come from GMO corn, but labels don't always say so unless the product is specifically marked 'organic' or 'non-GMO verified,'" Ms. Rodriguez explained. "Other countries require GMO labeling. It's an ongoing conversation about transparency and consumer choice."

"That seems like something people should know about," Ana said quietly.

"Many people agree with you," Ms. Rodriguez replied. "That's why reading labels, asking questions, and understanding where our food comes from is so important."

The Many Uses of Corn Products

Before leaving, Ms. Rodriguez showed them a display of everyday products that contained corn-derived ingredients. The kids were amazed:

- Foods: Cereal, bread, crackers, candy, soda, salad dressing, yogurt, ice cream

- Personal care: Toothpaste, shampoo, lotion, makeup

- Household: Cleaning products, laundry detergent, paper products

- Industrial: Adhesives, bioplastics, ethanol fuel

"Corn is one of the most versatile crops in the world," Ms. Rodriguez said. "It feeds people, feeds animals, powers cars, and creates thousands of products. That's why understanding how it's grown and processed matters."

As the bus pulled away from the plant, Ana looked at her notes. "I had no idea how complicated processing was. And how many products depend on corn."

Mary added, "It makes me think about what we eat. If corn is in so many things, and if there are questions about herbicides or GMOs, then learning about this stuff is really important."

Ben grinned. "Our science fair project is going to be epic."

That evening, Mary sketched a detailed diagram for the project: seed → field → harvest → processing → food product. She added notes at each step about where chemicals might remain, how processing removed or concentrated certain components, and where questions still existed.

Chapter 7 - Pantry Detectives and a Breakfast Clue

One morning Ana's household gathered for breakfast. Her mother, Rosa, sipped coffee and reached for toast. Ana noticed Rosa press her hand to her belly and breathe slowly.

"Are you okay, Mama?" Ana asked.

Rosa nodded. "Just a little stomach ache. It happens sometimes when I eat certain things. Don't worry."

But Ana did worry. She'd noticed her mother had been having more stomach troubles lately. That afternoon, while Rosa was at work, Ana decided to investigate.

She opened the pantry and pulled out item after item, reading labels carefully. The first thing she noticed: corn was everywhere.

Ana's Pantry Inventory

- Breakfast cereals: Listed corn meal, corn starch, and corn syrup

- Bread: Contained high-fructose corn syrup and cornstarch

- Canned soups: Had corn syrup, modified corn starch, and maltodextrin

- Salad dressing: Listed corn oil and corn syrup solids

- Snack bars: Contained corn syrup, corn flour, and dextrose (from > corn)

15

- Soda: High-fructose corn syrup was the second ingredient

- Tortilla chips: Made from corn (expected), but also had corn oil

- Yogurt: Contained modified corn starch as a thickener

- Even vitamins: Had maltodextrin (derived from corn) as a filler

Ana pulled out her notebook and made a chart with columns: "Product Name," "Corn Ingredients," and "When Mom Ate It." She started recording which foods her mother ate most often and when she complained about stomach problems.

Over the next week, Ana noticed a pattern. On days when Rosa ate a lot of processed foods for breakfast and lunch—cereal, bread, canned soup, snack bars—she was more likely to have stomach discomfort in the evening. On days when she ate simpler foods like eggs, fresh vegetables, rice, and fruit, she seemed to feel better.

"Correlation doesn't mean causation," Ana reminded herself, remembering Ms. Patel's science lessons. "But it's worth investigating."

She shared her findings with Mary and Ben at lunch the next day.

The Corn Crew Investigates Food Labels

Mary pulled out her own list. "I checked our pantry too. We have corn ingredients in probably 70% of the packaged foods."

Ben nodded. "Same at my house. My mom said she tries to buy less processed food, but it's hard because it's more expensive and takes longer to prepare."

The three friends decided to do a deeper investigation. They would:

1. Document all the different names for corn-derived ingredients.
2. Count how many foods in a typical grocery store contained corn.
3. Research whether corn-derived ingredients could affect health.
4. Interview Dr. Lopez about gut health and food additives.

That weekend, they met at Mary's house with stacks of empty food packages her family had saved for recycling. They spread them across the kitchen table and started reading labels.

Common Corn-Derived Ingredients

The kids were surprised to learn how many different names corn could hide under:

- Sweeteners: Corn syrup, high-fructose corn syrup (HFCS), dextrose, maltose, maltodextrin, glucose syrup

- Starches: Corn starch, modified corn starch, modified food starch (often from corn)

- Oils and fats: Corn oil, vegetable oil (often includes corn)

- Acids and alcohols: Citric acid (often corn-derived), xanthan gum (fermented from corn), sorbitol

- Proteins: Corn gluten, hydrolyzed corn protein

- Others: Corn meal, corn flour, corn bran, polenta, masa

"Wow," Ben said, staring at the list. "It's in everything."

Mary had been researching on her tablet. "It says here that corn is cheap and versatile. After World War II, industrial farming made corn production explode. Food companies started using corn derivatives instead of more expensive ingredients like cane sugar or butter."

Ana looked thoughtful. "But just because something is cheap and common doesn't mean it's the best choice for health."

They decided to add a section to their science fair project showing:
• How many corn-derived ingredients exist
• How often they appear in processed foods
• Why food manufacturers use them
• What health researchers say about consuming large amounts of these ingredients
Ana carefully organized her observations about her mother's symptoms. She made sure not to jump to conclusions, but she wanted Dr. Lopez to see the pattern she'd noticed.
"My mom has an appointment next week," Ben said. "Dr. Lopez said we could come and ask questions for our project."

The three friends felt like real scientists—gathering data, forming hypotheses, and preparing to test their ideas with expert help.

Chapter 8 - Talking with Dr. Lopez

Dr. Lopez welcomed the kids and their families into her office the following Wednesday afternoon. Her office was warm and friendly, with colorful anatomy posters on the walls and a skeleton model named "Bones" in the corner.

"You're asking really smart questions," Dr. Lopez began, settling into her chair. "Ana, your mom told me about your observations. And Ben mentioned you're doing a science fair project about corn and food processing."

Ana opened her notebook nervously. "I noticed my mom has stomach problems more often on days when she eats a lot of processed food. Could corn ingredients be causing it?"

Dr. Lopez smiled gently. "That's excellent detective work, Ana. Let me explain what we know—and what we're still learning."

Understanding the Gut Microbiome

"Inside our intestines, we have trillions of bacteria, fungi, and other microorganisms," Dr. Lopez explained. "We call this the gut microbiome. These microbes help us digest food, make vitamins, and protect us from harmful bacteria."

She showed them a colorful diagram of the digestive system. "A healthy microbiome has lots of diversity—many different types of beneficial bacteria. When that balance gets disrupted, it can cause digestive problems, inflammation, and even affect our immune system and mood."

Ben leaned forward. "What disrupts the balance?"

"Good question! Several things can affect the gut microbiome:

• Antibiotics (kill both bad and good bacteria)

• Diet (especially low fiber, high sugar, highly processed foods)

• Stress

• Environmental chemicals

• Lack of sleep"

Mary asked, "Could herbicides like glyphosate affect gut bacteria?"

Dr. Lopez nodded seriously. "There are laboratory studies showing that glyphosate can affect certain bacteria. It works by blocking an enzyme pathway called the shikimate pathway, which plants use to make amino acids. Humans don't have this pathway, but many gut bacteria do."

"So it could affect our gut bacteria?" Ana asked.

"Potentially, yes. But—and this is important—the amounts of glyphosate residues typically found in food are very small. We don't have clear evidence yet that these low levels significantly disrupt human gut microbiomes in real-world conditions. Scientists are still researching this."

The Problem with Ultra-Processed Foods

"Now, let me tell you about something we DO have strong evidence for," Dr. Lopez continued. "Ultra-processed foods."

She pulled out a chart showing different levels of food processing:

1. Unprocessed or minimally processed: Fresh fruits, vegetables, grains, meat, milk

2. Processed ingredients: Oil, butter, sugar, salt

3. Processed foods: Canned vegetables, cheese, fresh bread

4. Ultra-processed foods: Soda, chips, candy, instant noodles, packaged snacks, many breakfast cereals

"Ultra-processed foods are made mostly from substances extracted from foods—like oils, fats, sugar, starch, and protein—plus additives like flavors, colors, and preservatives," Dr. Lopez explained.

Ana looked at her list of pantry items. "That sounds like most of what we found."

"Exactly. And here's the concern: Studies from around the world show that people who eat a lot of ultra-processed foods have higher rates of:

• Obesity

• Type 2 diabetes

• Heart disease

• Some cancers

• Digestive problem

• Depression and anxiety"

Ben wrote quickly. "But why? Is it the specific ingredients, or something else?"

"Great question! Scientists think it's a combination of factors:

1. High in sugar, salt, and unhealthy fats

2. Low in fiber and nutrients

3. Designed to be hyperpalatable—so tasty that you want to eat more

4. Often eaten quickly, leading to overconsumption

5. May contain additives that affect gut bacteria or metabolism

6. Replace whole foods that would provide beneficial nutrients"

Mary raised her hand. "I actually tried an experiment on myself last week. I ate whole-food snacks instead of packaged ones for five days."

"What did you notice?" Dr. Lopez asked.

"My energy felt more stable. I didn't get hungry as fast. And I could focus better in class," Mary said thoughtfully.

"That's a perfect example of listening to your body," Dr. Lopez said. "Whole foods—like fruits, vegetables, whole grains, nuts, beans—contain fiber, vitamins, minerals, and phytonutrients that our bodies need. They also feed good gut bacteria."

What About Rosa's Symptoms?

Dr. Lopez turned to Ana. "Your observations about your mother are valuable. Based on what you've told me, I'd recommend she try a simple experiment: for two weeks, reduce ultra-processed foods and focus on whole foods. Keep a food and symptom diary."

Ana's eyes lit up. "Like a real science experiment!"

"Exactly! We call it an elimination trial. If her symptoms improve, we might have found a dietary trigger. It could be:
• Too much added sugar causing inflammation
• Low fiber leading to digestive issues
• Food additives she's sensitive to
• Lack of nutrients her gut needs
• Or possibly small amounts of pesticide residues, though that's less common"

Dr. Lopez continued, "For some people, eating fermented foods like yogurt, sauerkraut, or kefir can help restore gut bacteria. Fiber-rich foods like vegetables, fruits, whole grains, and beans feed the good bacteria."

Rosa, who had been listening quietly, spoke up. "I'm willing to try. I want to feel better."

Making Practical Changes

"I know changing your diet can be challenging," Dr. Lopez said. "Processed foods are convenient and affordable. But here are some realistic steps families can take:"

1. Start small: Replace one processed snack per day with a whole food option (apple slices, carrots, nuts)

2. Read labels: Choose products with fewer ingredients and ones you recognize

3. Cook more: Even simple cooking (scrambled eggs, rice and beans, roasted vegetables) is better than packaged meals

4. Buy frozen: Frozen fruits and vegetables are nutritious, affordable, and convenient

5. Plan ahead: Prep simple meals on weekends

6. Don't aim for perfection: An 80/20 approach (mostly whole foods, some processed) is realistic

Mary asked one more question. "If someone wants to avoid herbicide residues, should they buy organic?"

"Organic foods are grown without synthetic pesticides or herbicides," Dr. Lopez explained. "Testing shows they have lower residues. However, organic food is more expensive, and most conventional foods are within legal safety limits. It's a personal choice based on your values, budget, and health concerns."

She added, "The most important thing is to eat more fruits and vegetables—organic or conventional. The health benefits of eating produce far outweigh the small risks from residues."

Ben summarized in his notebook: "Biggest health issue, too many ultra-processed foods. Herbicide residues- smaller concern but worth learning about. Solution = eat more whole foods, read labels, make informed choices."

Dr. Lopez smiled. "You've got it. You kids are doing excellent science work. I'm proud of you."

As they left the office, Ana felt hopeful. She had data, expert advice, and a plan to help her mom feel better. Mary and Ben had valuable information for their science fair project. And all three understood that good health was about looking at the big picture—not just one ingredient or one chemical, but the whole food system and how it affected people's bodies.

That evening, Mary added a new section to their project: "Ultra-Processed Foods and Health." She included Dr. Lopez's advice, scientific studies about diet and disease, and practical tips for families.

The Corn Crew was beginning to see that their project wasn't just about corn or herbicides—it was about helping people make better choices for their health.

Chapter 9 - The Grocery Store Expedition

On a rainy Saturday, the three friends pushed a shopping cart through Miller's Grocery with determination. They had permission from the store manager to take notes and photos for their science fair project.

Ben carried his notebook. Mary had her camera. Ana had a clipboard with a check list of products.

"Let's be systematic," Mary said. "We'll go aisle by aisle and count products with corn-derived ingredients."

They started in the cereal aisle. Out of 47 cereals, 42 contained at least one corn ingredient. Most had multiple corn ingredients.

Ana made a note on the clipboard. "This one has corn meal, corn syrup, corn starch, AND yellow corn flour."

In the bread aisle: 31 out of 38 products contained corn ingredients, usually high-fructose corn syrup or cornstarch.

In the snack aisle: Nearly every product contained corn—corn chips, corn oil in crackers, corn syrup in granola bars.

In the beverage aisle, Ben counted: "18 out of 20 sodas have high-fructose corn syrup. Even some of the juice drinks do."

The condiments were full of corn: ketchup, barbecue sauce, salad dressings, mayonnaise. Even pickles listed corn syrup.

"This is wild," Mary said, photographing a shelf. "Corn is in practically everything."

The Data

After two hours, they compiled their results:

- Cereals: 89% contained corn ingredients

- Bread and baked goods: 82% contained corn ingredients

- Snacks (chips, crackers, bars): 95% contained corn ingredients

- Beverages (soda, juice drinks): 90% of sweetened drinks contained corn syrup

- Condiments and sauces: 78% contained corn ingredients

- Frozen meals: 71% contained corn ingredients

- Candy: 87% contained corn syrup

- Dairy products: 35% contained corn ingredients (mostly in sweetened yogurts)

"Now let's compare organic and conventional," Mary suggested.

They noticed that organic products:
- Cost 20-50% more on average
- Had fewer corn-derived ingredients (many used cane sugar instead of corn syrup)
- Had simpler ingredient lists
- Were less available (smaller selection)

Ben did quick math. "If a family switched entirely to organic, their grocery bill could go up $50-100 per week."

"That's a lot," Ana said. "Not everyone can afford that."

"Maybe the answer isn't all-or-nothing," Mary suggested. "People could choose organic for items where they're most concerned, and conventional for others. Or just focus on buying more whole foods that don't need labels— like fresh produce."

After the grocery store, Ben's mom took them to a fast-food restaurant for lunch. The kids applied their new knowledge.

"Look at the ingredients posted online," Mary said, scrolling on her phone. "The bun has high-fructose corn syrup. The ketchup has corn syrup. The chicken nuggets are breaded with cornstarch and corn flour. Even the soda is HFCS."

"Almost everything here contains corn or corn derivatives," Ben observed.

Ana bit into her chicken sandwich. "It tastes good, but now I'm thinking about all the processing that went into it. And whether there might be herbicide residues."

"That's called being an informed consumer," Ben's mom said. "Knowledge helps you make choices. You don't have to avoid everything processed, but you can be aware and make better decisions when possible."

That evening, Mary created a large chart for their science fair display showing the percentage of products containing corn ingredients across different food categories. Ana designed an infographic showing how much corn-derived food the average American consumed per year (estimated at over 50 pounds). Ben wrote up their observations about organic vs. conventional options.

The Corn Crew decided to include a section in their project called "Making Informed Choices" with practical tips:
• Read ingredient labels

• Choose minimally processed foods when possible

• Buy organic for items where you want to avoid pesticides (if affordable)

• Focus on whole foods (fresh produce, whole grains, beans, nuts)

• Don't stress over perfection—small changes add up

Their project was taking shape: from the cornfield to the processing plant to the grocery store to people's health. They were connecting all the dots.

Chapter 10 - The Data Hunt

With two weeks left before the science fair, the Corn Crew dove into research. They visited the town library and used school computers to find scientific articles, government reports, and news stories.
The more they read, the more they learned about the complex world of agricultural chemicals.

Learning the Language of Science

They learned new words that sounded dull at first, but each had a tiny door to a new idea:

Residue: A small amount of a chemical that stays behind after spraying

Metabolite, (pronounced "meh-TAB-uh-light."),: What chemicals become when they break down in the environment or in living things

Half-life: How long a chemical takes to fall to half its original amount

Maximum Residue Limit (MRL): The highest amount of a pesticide legally allowed to remain in food

Bioaccumulation: When a chemical builds up in an organism over time (happens with fat-soluble chemicals)

Water-soluble vs. fat-soluble: Water-soluble chemicals (like glyphosate) wash away more easily; fat-soluble chemicals stick around longer in soil and living tissues.

Understanding Glyphosate

Ben discovered that glyphosate, a common herbicide Uncle Ray used, worked by stopping a plant enzyme called EPSPS. Without that enzyme, plants couldn't make certain amino acids and would stop growing.

"But genetically modified corn has been engineered to tolerate glyphosate," Ben explained to his friends. "The corn has a different version of the EPSPS enzyme that still works even when glyphosate is present."

"So the weeds die, but the corn survives," Mary said.

"Exactly. That's why it's so popular with farmers."

They learned that other herbicides, like atrazine,(pronounced "a-truh-zeen."), worked differently. Atrazine was less water-soluble and stuck to soil particles longer. It could contaminate groundwater and had been banned in Europe but was still used in the United States.

What the Research Shows

The kids visited the town library and read reports from scientific journals, with help from the librarian, Ms. Chen, who showed them how to find reliable sources. They learned that:
Government agencies like the EPA and FDA set safety limits based on studies in lab animals.

- Most food samples tested show residues below these legal limits

- Independent scientists have found glyphosate in many foods, drinking water, and even in people's urine

- The amounts detected are usually very small and come and go from the body within days

- However, repeated exposure means small amounts may be present almost constantly

- Some lab studies suggest glyphosate and other herbicides might disrupt hormones, affect gut bacteria, or cause other health effects at low doses

- Large population studies show mixed results—some find associations with health problems, others don't

- The scientific community is debating what these findings mean for real-world human health.

27

"So the science isn't settled?" Ana asked Ms. Chen.

"That's right," the librarian said. "Science rarely gives simple yes-or-no answers quickly. Different studies use different methods and sometimes get different results. That's why scientists keep researching and debating."

Mary wrote in her notebook: "Important to present both sides in our project. Show what we know, what we don't know, and what questions remain."

Observing the Local Environment

The Corn Crew also did fieldwork around Uncle Ray's farm. They wanted to see if they could observe any environmental differences between sprayed and unsprayed areas.

With Uncle Ray's permission and supervision, they compared two patches:

Patch A: Regularly sprayed with herbicides
Patch B: An unsprayed border area with wildflowers

They spent three afternoons observing and counting:

- Insects: Patch B had 3 times more butterflies, bees, and beneficial beetles

- Plants: Patch B had 12 different plant species; Patch A had only corn and a few stubborn weeds

- Soil: Patch B soil was darker and crumblier; Patch A soil was lighter and firmer

- Earthworms: They found 8 worms in a shovelful from Patch B, only 2 in Patch A

"This is interesting data," Uncle Ray said, looking at their observations. "But remember, I chose not to spray Patch B deliberately to create habitat for pollinators. The differences you see are partly because of the spraying, but also because Patch B has wildflowers that attract insects."

"So we have to be careful about causation," Mary said.

"Exactly. Correlation doesn't prove causation. But it does suggest herbicides affect more than just weeds. The biodiversity is different."

Ben added this to their project: "Herbicides can affect farm ecosystems, reducing plant and insect diversity. This might impact pollinators, natural pest control, and soil health."

The kids also learned about alternative farming practices:

• Cover cropping: Planting crops like clover or rye between growing seasons to prevent weeds and improve soil
• Crop rotation: Changing what's planted each year to break weed and pest cycles
• No-till farming: Not plowing the soil to preserve structure and reduce erosion
• Integrated pest management: Using multiple methods (beneficial insects, targeted spraying, resistant varieties) instead of relying only on chemicals
• Organic farming: Avoiding synthetic pesticides and fertilizers

Uncle Ray explained that these methods had tradeoffs: they could be more labor-intensive, might produce lower yields, and weren't always economically feasible for large operations. But they offered environmental benefits.

"The important thing," Uncle Ray said, "is that farmers, scientists, and consumers keep talking and learning. Agriculture has to feed billions of people while protecting the environment. That's not easy."

The Corn Crew felt the weight of complexity. There were no simple answers, but asking good questions was a start.

Chapter 11 - A Small Experiment

With one week left, the Corn Crew examined their experiments for their science fair display. They wanted hands-on demonstrations that other kids could learn from.

Experiment 1: Plant Growth Comparison

Mary, Ana, and Ben planted three small pots with the same soil from Uncle Ray's farm. They used bean seeds because they sprouted quickly.

Pot 1 (Control): Regular soil, water, sunlight
Pot 2 (Compost): Soil with added compost, water, sunlight
Pot 3 (Herbicide): Soil with a dose of diluted herbicide to mimic spray drift, water, sunlight

Over three weeks, they measured sprout height every three days, photographed leaf color, and counted insect visitors.

Results:

- Pot 2 (compost) grew the tallest and fastest—leaves were dark green and healthy
  Pot 1 (control) grew well, with medium-green leaves
  Pot 3 (herbicide) grew slower—leaves were paler and smaller, and one seedling didn't survive
  Insect observations: Pot 2 attracted small beetles; Pot 3 had no visible insect visitors.
  "Our experiment was small and not conclusive," Mary wrote in the project report. "But it shows that herbicides can affect plant health even at low doses, and may reduce beneficial insects around treated areas."

30

Experiment 2: Composting for Soil Health

At school, the Corn Crew helped set up a composting bin. Students contributed fruit peels, vegetable scraps, leaves, and shredded paper.

Within weeks, the pile heated up as microbes broke down the organic matter. After a month, they had dark, crumbly compost.

They tested the compost by planting identical bean seeds in regular soil and compost-enriched soil. The compost plants grew noticeably faster and attracted more earthworms.

"Composting is a great way to reduce food waste and create healthy soil without chemicals," Ana explained in their project notes.

Experiment 3: Testing Soil pH

The kids tested soil pH using simple household materials: vinegar (acid) and baking soda (base).

They collected soil samples from different areas around Uncle Ray's farm and their own yards. When they added vinegar to soil, fizzing indicated alkaline soil. When they added baking soda mixed with water, fizzing indicated acidic soil.

They found that soils near sprayed areas tended to be slightly less acidic than unsprayed wooded areas. Uncle Ray explained that some fertilizers and herbicides can change soil pH over time, which affects which plants and microbes can thrive.

Experiment 4: Microbial Activity Jars

For their science fair display, the kids created two jars with identical vegetable scraps and water.

Jar 1: Water from a clean upstream location
Jar 2: Water with a tiny, documented amount of soil extract from near the farm drainage area

Over several days, Jar 1 showed more bubbling and faster decomposition (signs of microbial activity). Jar 2 had less visible activity.

"This is a simplified model," Ben wrote. "It suggests that environmental chemicals can affect microorganisms that break down organic matter. Microorganisms are important for soil health and gut health."

They were careful to note: "This is not direct evidence of human health effects. It's a demonstration of how chemicals can affect microbial systems."

The experiments gave the Corn Crew concrete demonstrations for their science fair project. They weren't trying to prove that herbicides were dangerous—they were showing how farming practices affect plants, soil, insects, and microbes in observable ways.

Their project was about curiosity, careful observation, and making informed decisions based on evidence.

Chapter 12 - The Last Weekend: Interviews and Final Checks

With one weekend left before the science fair, the team moved into high gear.

They had data from the grocery store, photos from the farm and processing plant, results from their experiments, and notes from Dr. Lopez. Now they needed to organize everything into a clear, compelling presentation.

Gathering Expert Perspectives

Mary's mother helped arrange for several people to be available for questions at the science fair:

• Uncle Ray: Would talk about farming practices, chemical use, and economic pressures
• Dr. Lopez: Would explain gut health, ultra-processed foods, and current research
• Ms. Rodriguez from the processing plant: Would answer questions about industrial food processing
• Mr. Chen, an organic farmer from the next county: Would discuss organic and regenerative farming methods
• Ms. Park, a nutritionist: Would talk about healthy eating on a budget

"We want to show different perspectives," Mary explained. "Not everyone agrees about the best way to farm or eat. That's okay. People can hear different viewpoints and make their own decisions.

Creating the Display

The team designed their science fair display on three large poster boards that would stand side by side. They worked at Mary's dining room table, carefully arranging printouts, charts, and photos.

LEFT BOARD: From Seed to Table

- • Flowchart showing: Seed → Planting → Growing → Spraying → Harvest → Processing → Grocery Store → Your Plate

- • Photos of farm equipment and processing plant

- • Explanation of GMO corn and herbicide tolerance

- • What are herbicides and how do they work?

- • Maximum Residue Limits (MRLs) and food safety testing

CENTER BOARD: Research and Experiments

- • Grocery store data (chart showing % of foods with corn ingredients by category)

- • Results from plant growth experiment (photos and measurements)

- • Soil pH testing results

- • Microbial activity jar observations

- • Quotes from Dr. Lopez about gut health

- • Scientific vocabulary (residue, metabolite, half-life, MRL, bioaccumulation)

RIGHT BOARD: Health, Choices, and Action

- • Ultra-processed foods and health risks (chart showing disease associations)

- • Ana's observations about her mother's symptoms (presented as a case study)

- • Tips for making healthier food choices

- • Organic vs. conventional: pros, cons, and costs

- • Alternative farming methods (cover cropping, crop rotation, organic)

- • Section titled 'What We Don't Know (Yet)' with unanswered questions

At the top of the center board, they added a bold title:

"From Seed to Spoon: Corn, Chemicals, and Our Choices"

They practiced their presentation over and over, timing themselves to make sure they could explain everything in 10 minutes.

Ben summarized their key message: "Corn is in a huge number of foods. It's grown with herbicides that can leave small residues. The biggest health concern isn't necessarily the residues—it's that too many corn-derived ingredients appear in ultra-processed foods that aren't good for us. By learning about the food system, we can make better choices."

Ana added, "And there are lots of choices to make at every level—farmers choosing how to grow, companies choosing how to process, families choosing what to eat. Everyone has power."

Mary smiled. "Let's do this."

Chapter 13 - Science Fair Day

The gym smelled of glue sticks on the morning of the science fair. Posters leaned against tables. Volcano models stood beside robotics projects and displays about weather patterns.

Mary, Ana, and Ben's display was positioned near the center of the gym. Their three poster boards stood proudly, covered with photos, charts, and carefully organized information.

The title read in bold letters: "From Seed to Spoon: Corn, Chemicals, and Our Choices"

The kids wore matching t-shirts that said "Corn Crew" with cartoon corn stalks. Ben's mom had made them as a surprise.

Students and parents began filing in. The judging would start at 10 AM, followed by open viewing for families and the community.

The Presentation

At 10:15, three judges approached their table. The kids took deep breaths and began.

Mary started: "We followed corn from seed to harvest to processing to the grocery store and finally to people's health. We learned that corn is in hundreds of products—not just obvious ones like corn tortillas, but also in processed foods as corn syrup, cornstarch, corn oil, and other derivatives."

She gestured to the left board. "Some corn is genetically modified to tolerate herbicides like glyphosate. This helps farmers control weeds, but it raises questions about chemical residues in food and environmental impacts."

Ana pointed to the center board. "We did several experiments. We planted seeds in soil with and without herbicide to observe growth differences. We tested soil pH from different areas. We created jars to model how chemicals might affect microorganisms."

"We also collected data," Ben added, pointing to the grocery store chart. "We found that 70-90% of processed foods in most categories contain corn-derived ingredients. which affects a huge portion of what Americans eat." Especially if there are concerns about herbicide residues or ultra-processed foods,

Mary continued, "We interviewed Dr. Lopez, a family physician. She told us that the biggest health concern isn't necessarily small herbicide residues—although scientists are still studying that—but rather the high consumption of ultra-processed foods."

Ana explained, "Ultra-processed foods are high in sugar, salt, and unhealthy fats, but low in fiber and nutrients. Studies link them to obesity, diabetes, heart disease, and other health problems."

She showed the case study timeline. "My mother has been having stomach issues. When she reduced ultra-processed foods and ate more whole foods, her symptoms improved. This is just one example, but it matches what research shows about diet and health."

Ben pointed to the right board. "We learned that people have choices at every level. Farmers can choose conventional or organic methods, cover cropping, or crop rotation. Food companies can use fewer additives and less processing. Families can read labels and buy more whole foods. Kids can ask questions and learn about where their food comes from."

Mary concluded, "We titled one section 'What We Don't Know (Yet)' because science doesn't have all the answers. Questions remain about long-term, low-dose exposure to herbicides, about how processing changes food, and about the best ways to feed the world while protecting the environment. But asking questions is how we learn."

The Questions

The judges asked sharp questions:

"How did you control variables in your plant experiment?"
"We used seeds from the same packet, identical soil from the same location, the same amount of water and sunlight, and the same size pots. The only difference was what we added—compost or herbicide."

"How do you know the health effects you mentioned are caused by ultra-processed foods and not by other factors?"
"We don't know for certain. Scientists use large studies that control for many factors like age, exercise, and smoking. They find associations between ultra-processed food consumption and health problems even after accounting for other factors. But correlation doesn't prove causation—that's why we're careful about how we present the evidence."

"Did you consider the benefits of modern agriculture, like feeding a growing population?"

"Yes! Uncle Ray explained that herbicides and GMO crops help farmers produce more food efficiently. Without modern farming methods, food would be more expensive and scarce. We tried to present both benefits and concerns in our project."

"What would you recommend to families based on your research?"
"Read labels, cook more with whole foods when possible, buy organic if it fits your budget and values, support local farmers, and don't stress about perfection. Small changes add up."

The judges nodded, made notes, and moved to the next project.

Community Conversations

After judging, the fair opened to the public. Families, teachers, and community members walked through the gym, stopping to look at projects.

The Corn Crew's display attracted a crowd. Uncle Ray came and answered questions about farming. Dr. Lopez talked to parents about nutrition. Ms. Rodriguez from the processing plant explained how industrial food production worked.

Mr. Chen, the organic farmer, stood beside Uncle Ray. The two had different approaches to farming, but they respected each other and enjoyed the discussion.

One parent asked, "Should I only buy organic food?"

Mr. Chen answered, "Organic has benefits—fewer synthetic pesticides, better for the environment in some ways. But it's more expensive. If you can't afford all organic, don't worry. Eating more fruits and vegetables—organic or conventional—is more important than avoiding all pesticides."

Uncle Ray added, "I use herbicides carefully and follow all safety regulations. I'm also trying cover crops on some fields to reduce erosion and improve soil. Farmers are always learning and adapting."

A grandmother stopped to read about ultra-processed foods. "This explains why my grandchildren are so picky and always want packaged snacks. They're designed to be irresistible."

Ms. Park, the nutritionist, nodded. "Ultra-processed foods are engineered to hit our pleasure centers—the perfect combination of sugar, salt, and fat. It's hard to compete with that when you're offering an apple. But kids' tastes can change if they're exposed to whole foods regularly."

The conversations continued all afternoon. People didn't always agree, but they listened and learned from each other.

The Results

At 3 PM, Ms. Patel gathered everyone for the awards ceremony.

The Corn Crew held hands nervously as categories were announced: Best Engineering Design, Most Creative, Best Use of Technology.

Finally: "Best Overall Group Project—for thorough research, careful experimentation, balanced presentation of complex issues, and community engagement—goes to Mary Johnson, Ana Rivera, and Ben Thompson for 'From Seed to Spoon: Corn, Chemicals, and Our Choices.'"
The gym erupted in applause.

The three friends hugged each other, grinning. They walked to the front to receive a blue ribbon and a certificate.

Ms. Patel smiled at them. "You did something really special. You didn't just answer a question—you started a conversation. That's what good science does."

Later, as families packed up projects and headed home, Ana looked at the blue ribbon in her hand. "We did it."

Mary nodded. "And I think we actually helped people learn something important."

Ben closed his notebook with satisfaction. "So what are we investigating next?"
They laughed. The science fair was over, but their curiosity wasn't.

The End

Epilogue - Seeds of Curiosity

Months later, the three of them stood at the edge of Uncle Ray's field again. The corn had been harvested; the rows were quiet. Stubble poked through the soil, and a light snow dusted the ground.

The world around them hadn't changed overnight, but their view of it had.

"We did a lot in four weeks," Ben said, tucking his notebook under his arm.

"We started with a maze and a map," Mary replied.

Ana smiled. "And we learned how to ask the right questions."

They weren't scientists yet—just three curious kids with notebooks and a willingness to learn. But curiosity was the first step.

Their project had been an invitation: to families, schools, and neighbors to think critically, ask experts, listen to different perspectives, and make informed choices.

Uncle Ray waved from the farmhouse porch. "So what are you kids going to investigate next?"

Mary looked at Ana and Ben with a mischievous grin. "Maybe sugar? Or how much plastic is in the ocean?"

Ben pulled out his notebook, already scribbling. "Or maybe how much of a backyard garden you need to feed a family for a week. I bet my dad would help us build raised beds."

"Or pollinators," Ana suggested. "I read that bee populations are declining. Maybe we could plant a pollinator garden at school."

They laughed and started walking back through the field, their boots crunching on frozen stubble.

Mary, Ana, and Ben understood that systems were complicated. They knew that not every farmer used harmful practices, that not every factory-produced food was terrible, and that there were tradeoffs in every decision.

What mattered was asking questions, testing ideas, listening to experts, respecting different viewpoints, and working toward solutions that balanced human needs with environmental health.

They carried their notebooks, their questions, and the small, steady confidence that comes from learning how to learn.

The cornfield maze had taught them something important: every path leads somewhere, but the best journeys start with curiosity and a willingness to explore.

And so, the Corn Crew headed home—already planning their next adventure.

Appendix - Quick Facts for Older Readers

Key Concepts and Definitions

• Soil Ecology: Soil contains billions of microorganisms per teaspoon—bacteria, fungi, protozoa, and more. These organisms decompose organic matter, cycle nutrients, and help plants grow. Healthy soil is essential for food production.

• Genetically Modified Organisms (GMOs): Organisms whose DNA has been altered using genetic engineering. Common GMO crops include corn, soybeans, cotton, and canola. GMO corn may be engineered to tolerate herbicides (like glyphosate) or to produce proteins that kill insect pests (like Bt corn).

• Herbicides: Chemicals used to kill unwanted plants (weeds). Different herbicides work in different ways. Glyphosate blocks the EPSPS enzyme, preventing plants from making essential amino acids. GMO crops can tolerate glyphosate because they have a modified version of this enzyme.

• Residues and Maximum Residue Limits (MRLs): Residues are small amounts of pesticides that remain on or in food after application. Government agencies like the EPA and FDA set MRLs—the maximum amount legally allowed. These limits are based on studies and include safety margins.

• Water Solubility: How well a chemical dissolves in water. Water-soluble chemicals (like glyphosate) wash away more easily but can contaminate water supplies. Fat-soluble chemicals stick to soil particles and living tissues longer and can bioaccumulate.

• Half-Life: The time it takes for half of a chemical to break down in the environment. Glyphosate has a half-life of days to weeks in soil, depending on conditions. Some older pesticides (like DDT) have half-lives of years and persist in the environment.

• Metabolites: Substances formed when chemicals break down. Sometimes metabolites are more toxic than the original chemical. Scientists study both the parent chemical and its metabolites when assessing safety.

• Ultra-Processed Foods: Foods made mostly from industrially processed substances (oils, fats, sugars, starch, protein) with added flavors, colors, and preservatives. Examples include sodas, chips, instant noodles, packaged snacks, and many breakfast cereals. Research links high consumption of ultra-processed foods to obesity, diabetes, heart disease, and other health problems.

• Gut Microbiome: The community of trillions of microorganisms living in the human digestive system. A healthy, diverse microbiome helps with digestion, immunity, and even mental health. Diet strongly influences the microbiome—fiber and fermented foods support beneficial bacteria.

• Organic Farming: Farming without synthetic pesticides, herbicides, or fertilizers. Organic farmers use crop rotation, compost, cover crops, and biological pest control. Organic foods generally have lower pesticide residues but may be more expensive.

• Cover Cropping: Planting crops (like clover, rye, or vetch) during the off-season to prevent erosion, suppress weeds, improve soil health, and add nutrients. Cover crops are often tilled back into the soil before planting the main crop.

• Crop Rotation: Changing which crops are planted in a field each year. This breaks pest and disease cycles, improves soil health, and can reduce the need for pesticides and fertilizers.

Important Reminders

• Science is rarely black and white. Most questions have complex answers that depend on many factors.

• Correlation does not equal causation. Just because two things happen together doesn't mean one caused the other.

• Different experts may have different opinions based on their training, research, and values. Listen to multiple perspectives.

• The best choices balance multiple goals: feeding people, protecting health, preserving the environment, and supporting farmers' livelihoods.

• Small actions matter. Reading labels, reducing processed foods, composting, and asking questions all make a difference.

• Keep learning! Science advances, and our understanding of food, agriculture, and health continues to evolve.

Further Reading and Resources

(Educators: See the Educator Guide for an extensive list of age-appropriate books, websites, videos, and activities.)